CHAIN OF D

JAMES BERRY

Chain of Days

Oxford New York
OXFORD UNIVERSITY PRESS
1985

Oxford University Press, Walton Street, Oxford OX2 6DP
London New York Toronto
Delhi Bombay Calcutta Madras Karachi
Kuala Lumpur Singapore Hong Kong Tokyo
Nairobi Dar es Salaam Cape Town
Melbourne Auckland
and associated companies in
Beirut Berlin Ibadan Mexico City Nicosia

Oxford is a trade mark of Oxford University Press

First published 1985 by Oxford University Press

British Library Cataloguing in Publication Data
Berry, James, 1924 –
Chain of days.
I. Title
821'.914 PR9265.9.B/
ISBN 0-19-211964-8

Set by Getset Ltd.
Printed in Great Britain by
Biddles Ltd.
Guildford, Surrey

For Mary

Acknowledgements

for poems here
that appeared in other publications
are due to:

Fiction Magazine for 'Chain of Days'

Litmus for 'Ongoing Encounter'

News for Babylon for 'Caribbean Proverb Poems', 'On an Afternoon Train from Purley to Victoria, 1955', 'New World Colonial Child', 'Confession', 'Thinking Loud-Loud', 'Speech for Alternative Creation'

Kunapipi for 'The Testing of the Ideal Third World New Man'

Bluefoot Traveller for 'Other Side of Town'

Outposts for 'Distance of a City'

Bananas for 'New World Black Woman'

South East Arts Review for 'Last Freedom of Martin Luther King'

Ambit for 'The Presence', 'Great Story', 'Approach and Response', 'Young Woman Reassesses'

Poetry Review for 'Fantasy of an African Boy', 'From Lucy: Englan' A University', 'A Companionship', 'Story by Bodyparts', 'It's Me Man', 'Detention and Departure', 'New Reading Like Rebellion', 'Thinking Back on Yard Time'

New Departures for 'In-a Brixtan Markit', 'I Am Racism'

Oxford Poetry for 'Letter to My Father from London'

Contents

Chain of Days

Night Comes too Soon

Here now skyline assembles fire.
The sun collects up to leave.
Its bright following paled,
suddenly all goes. Dusk rushes
in, like door closed on windowless room.
Children go a little sad.

Fowls come in ones and groups
and fly up with a cry
and settle, in warm air branches.
Tethered pigs are lounging
in dugout ground.

Muzzled goat kids make muffled
cries. Cows call calves locked away.
Last donkey-riders come homeward
calling 'Good night!'
Children go a little sad.

Knives-making from flattened
big nails must stop. Kite ribs
of tied sticks must not develop.
Half shapes growing into bats
and balls, into wheels and tops
must cease by night's veto.

And, alone on shelves, in clusters
on the ground in corners, on
underhouse ledges, these
lovable embryos
don't grow in sleeptime.
Children go a little sad.

Bats come out in swarms.
Oil lamps come up glowing
all through a palmtree village.
Everybody'll be indoors
like logs locked up.
Children go a little sad.

Caribbean Proverb Poems 1

Dog mornin prayer is, Laard
wha teday, a bone or a blow?

Tiger wahn fi nyam pickney, tiger sey
he could-a swear e woz puss.

If yu cahn mek plenty yeyewater
fi funeral, start a-bawl early mornin.

Caribbean Proverb Poems 2

1
Hungrybelly an Fullbelly
dohn walk same pass.

Fullbelly always a-tell Emptybelly
'Keep heart'.

2
Yu fraid fi yeye
yu cahn nyam cowhead.

Yeye meet yeye
an man fraid!

3
Yu si yu neighbour beard
ketch fire, yu tek water
an wet fi yu.

When lonely man dead,
grass come grow a him door.

4
Satan may be ol
but Satan not bedriddn.

Man who is all honey,
fly dem goin nyam him up.

nyam: eat *pickney:* child or children *yeye:* eye

Chain of Days

In the spicing the salting and the blackening
I'm poling up in fires of summerlight.
A tree blooms from my umbilical cord.
I look for a touch from every eye.
Darkness the wide shawl with sun's heat,
my mother's songs go from wooden walls.
Humming birds hovering
hibiscus nodding
I dance in the eyes of an open house.

On a summer road under swinging palms
a chain of days showed me bewildered.

Taking the drumming of the sea on land
I take the rooted gestures,
I take the caged growing.
Little lamps merely tarnish
a whole night's darkness.
Our stories bring out Rolling-Calves
with eyes of fire and trail of long chain.
I look for a sign in every face,
particularly in my father's face.

A joy is trapped in me.
Voices of rainbow birds shower my head.
I decapitate my naked toe.
My eye traps warm dust.
I'm born to trot about delivering
people's words and exchanged favours
or just grains of corn or sugar or salt.

Scooped water dances
in the bucket perched on me.
My bellows of breath make leaps of flames.
I muzzle goat kids in raging sunsets
and take their mother's milk in morning light.

Rolling-Calves: monstrous evil spirits, in the form of calves, who haunt the countryside at night

Beaten there to remember I know nothing
I run to school with a page of book
and clean toenails and teeth.

All my homestead eggs go to market.
Farthings are the wheels that work my world.
My needs immaterial, I know I'm alien.

My father stutters before authority.
His speeches have no important listener.
No idea that operates my father
invites me to approach him.
And I wash my father's feet in sunset
in a wooden bowl.

And my father's toothless mother is wise.
My father coaxes half dead cows
and horses and roots in the ground;
he whispers to them sweetly and shows
the fine animal coat he persuaded to shine;
he pulls prized yams from soft earth
like big babies at birth
to be carried to rich tables.

Confused and lonely I sulk.
Companioned and lost I laugh.
I fill hunger with games.

> On a summer road under swinging palms
> a chain of days showed me bewildered.

I go to the wood
it is a neighbour.
I go to the sea
it is a playground.
I circle track marked hills.
I circle feet cut flatlands.

I turn rocks I turn leaves.
I hunt stone and nut marbles in woods.
I rob pigeon pairs from high trees.
Wasps inflate my face for a soursop.
I beat big nails into knives.
Tops and wheels and balls come from wood.
Yet movement and shape mystify:
I am tantalized.

I understand I'm mistaken
to know I'm truly lovable, and that
my lovable people are truly lovable.
I understand something makes me alien.
I wonder why so many seedlings wither
like my father's words before authority.

Wondering, dreaming, overawed, I sit
in the little room with faces like mine.
Low lamp light spreads a study
of the past on our faces.
I sleep with the purple of berries
on my tongue, and I warm thin walls.

On a summer road under swinging palms
a chain of days showed me bewildered.

I keep back the pig's squeal
or the young ram's holler
when my father takes out their balls
with his own razor.
Water I pour washes my father's hands.
I wish my father would speak.
I wish my father would use
magic words I know he knows.
I wish my father would touch me.

Past winds have dumped
movements my ancestors made.
I dread my father's days
will boomerang on me.
I want to stop time and go with time.
In the hills alone I call to time.
My voice comes back in the trees
and wind. Why isn't there the idea
to offer me as sacrifice like Abraham's son?

My sister goes and washes
her breasts in the river.
Like a holy act I wash
in my brothers' dirty water.

On a summer road under swinging palms
a chain of days showed me bewildered.

My mother's dead granny brings
medicines to her in dreams.
My mother is a magician.
My mother knows how to ignore my father.
My mother puts food and clothes
together out of air. Bush and bark and grasses
work for my mother. She stops
the wickedest vomiting. She tells you
when you haven't got a headache at all.
In the pull of my mother's voice
and hands she stings and she washes me.

Put me in bright eye of sunlight
in shadows under broad hats,
on hillside pulling beans,
or chopping or planting,
I am restful with my mother.

The smell of sunny fields in my clothes
I meet my mother's newborn in the secret
birth room strong with asafetida.
I say goodbye touching my mother's mother
in the yard, her cold face
rigid toward the sky.

Quickfooted in the summer
I come over dust and rocks.
Echo after echo leaves me
from the edge of the sea,
from the edge of the sea.

Surprise Meeting in Port Antonio

Crown and airs and pole straight men
oppress me I fly
from my parish courthouse
and summer chills my brain

My old pal the swift head in class
the laughing fool for fun and work
the cricketer by birth
emasculated in law in his
inheritance of failure

Long necked he stood
underfed and passive
inside out his pockets bared
of three shillings
and ten pence hapenny

All the wealth in his name on earth
warm coins the law passed
to the giggling girl at school
turned sad faced skeleton mother
of six by three men
whose absence overwhelmed her room

Must rain through roofs drown us
Must sunlight consume us

Visitor Village Find

Look yah now
look in-a dis yah hut
pon chop-chop log

Granmumma
wid no back fi straightn up
no soun foot fi walk
no teet fi bite
an not one fowl a yard
not even a rat in-a house
or ants a-run roun cupbad

Only Granmumma wid three
mosquito-leg pickney dem

Granmumma
why mek yu so wid so-so self
wid bare sun-yeye a-watch
down fram house top?

Ow lang yu si-down yah
wid children dem wid so-so place?

Ow lang yu si-down yah
an nobody a-hear God a-sey
to come an touch yu —
yu an de pickney dem?

yah: here
yeye: eye
foot: leg
cupbad: cupboard
pickney: child or children

Detention and Departure

I shouldn't live, you know, man.
I shouldn't live,
too much hunger devoured me,
too many lacks hit me in
and stamped me down
all heels over head long long time.

I washed up here washed-up,
an overdone case, man,
rolled from ship of fire,
tongue clipped,
steps short.

A dry stick sun-charged man,
I powered axe,
powered crowbar and hoe,
I flattened hills.
I made mountains make valleys.
I banked back all of the sea this side.

Then turned into a bottomless bag, man,
I swallowed up all the sea-sound.
I saw I sucked in the force of the sea.

I lifted, man,
lifted up,
dropped off white bones like shell.

I climbed this high noon, man,
climbed this high noon,
hands sharp,
sharper than steel,
head big,
loaded with eyes.

I took this high noon, man,
took this high noon,
in claim of all earth,
in claim of all earth, man.

The Testing of the Ideal Third World New Man

(for Caribbean Leaders)

Arms raised he stirs exultation
beyond rotten roots that renewed him
the sad god of poverty who grew secretly
obsessed with strange identities
to risk the shocks of bluffs of blood
a man with thorny head of subjugation
who emerged from low lying hideaways
who grafted glorious eyes of success
and practised how to breathe in cash
how to move in it and gesture with it
and he drilled himself in other languages
drilled himself to outpace
and outdazzle conquerors
till every day reflected him
as the swordsharp magic machete
and he strung little victories together
like stepping stones
and this
upsurge of nonconforming
is all-real
and troubles establishments like a flier from Mars
and puts up new polished milestones
and his words are busy over ocean waves
his missions clothe naked absences
and empty banks into empty pockets
and the world knows him a fresh agitator
our eye and target of the nation

Listen
hear him announced: 'Here he is
knower of tracks to every fireside
sound sleeper on bare boards
good company on parched corn and water for dinner
the bringer
of new dimensions
new textures before the eyes
and bodies growing scarless
our sage
who brightens blackness
our floods on dry time
our builder after hurricane
our own man who talks 'roots talk'
as well as 'professor talk'
who finds the lost
who wakens the dead and all beginners
who is hard transparent glass
with deep reds and blues
our leader
here he is here he is
our man with eyes all round the head'

Fantasy of an African Boy

Just Being

I laughed and my echoes shook apples
off trees of a thousand lands

My swimming trailed one long lonely road
all deep in the sea

I stepped out and stood on a mountain top
high up in a sheet of sunlight

Dogwood Tree

Only now I begin to see you
a traveller,
mysterious Dogwood Tree.

Topping your stance here
dark green, beaten shaped,
you are a waver draped with vines
and rammed with limbs,
not tall, but managing to reach
for some glare of light.

No signs of treatment
show on you,
over leaf changes
and oncoming footsteps.

You have been sawed and chopped.
Parts gone as shafts of carts
and tools and fence posts that grow.

And you try to overcome
your stumps: I stroke
your straight wood.

I want to know.
I want to know how
you took your incentive
in delight in the sun's eye
and your unyielding character.

I want to know if you are
an impulse or an echo
that became this old iron tree
and medicine tree,
with a juice of bark to make
a toothache stop
and make fish rise
and float sideways.

Going With All-Time Song

  Say
Goodmorning Love
in the midnight our market

  Say
Yesterday showed
our yearbook of yearning

Today washes
the words and eyes
in wonderment

  Say
Tomorrow is the torch focussed
now is my noontime

in the waking
  singing singing
I AM IN LOVE

Fantasy of an African Boy

Such a peculiar lot
we are, we people
without money, in daylong
yearlong sunlight, knowing
money is somewhere, somewhere.

Everybody says it's a big
bigger brain bother now,
money. Such millions and millions
of us don't manage at all
without it, like war going on.

And we can't eat it. Yet
without it our heads alone
stay big, as lots and lots do,
coming from nowhere joyful,
going nowhere happy.

We can't drink it up. Yet
without it we shrivel when small
and stop forever
where we stopped,
as lots and lots do.

We can't read money for books.
Yet without it we don't
read, don't write numbers,
don't open gates in other countries,
as lots and lots never do.

We can't use money to bandage
sores, can't pound it
to powder for sick eyes
and sick bellies. Yet without
it, flesh melts from our bones.

Such walled-round gentlemen
overseas minding money! Such
bigtime gentlemen, body guarded
because of too much respect
and too many wishes on them.

too many wishes, everywhere,
wanting them to let go
magic of money, and let it fly
away, everywhere, day and night,
just like dropped leaves in wind!

In God's Greatest Country, 1945

In this Lake Okeechobee land
of hibiscus, oranges and flamingos,
grass could deceive
it was sugarcane.

New like a city boy in
deep woods, I stood inside
the back of the bus, watching
empty seats in front
marked WHITES ONLY.

My friend sat, as any man sits
in a vacant public seat,
and the sun was attacked.
Horns grew in faces.

And the lady squirmed.
She yelled her person's purity
is blotted: a Black
violates her side.

Passions braked the bus.
The driver stood correctly,
legally, holding unholstered gun
coolly, like a Bible
to convert a Black.

'I'm British', my friend said.
But under steel of eyes
there was a cooler confidence,
'Niggers are jes niggers.'

We stepped down between
fields of nodding sugarcane.
Pop-eyed, at the back of the bus,
with sheep caged faces,
the black Americans watched us go
across the country road.

In the free sunlight,
satisfying the other tribe,
we walked into the little
segregated town of Belleglade, Florida.

Notes on a Town on the Everglades, 1945

Sounds of 'Ma Baby Gone'
make the ghetto air in blues
in this Southern town.

It charges me with dread
with puzzlement with wonder
at this haunted mass of black people.

They hum the blues in shanty streets
and open fields. They dance the blues
in bars. They pray blues fashion.

They gesture and move and look
like refugees or campers on home ground,
or just a surplus of national propagation.

And fenced in, policed, blues-ridden,
the people plant wounds
on any close body it seems.

Women and men, all ages, go and come
in bandaged movements
like hospital escapees.

And on the compelling side,
the gleaming nearness of town,
the bridge is policing.

White men through their streets,
like white men in the fields,
are knowing and proud stalwarts.

With cold eyes like passionless gods
their groomed bodies go
extended with guns.

Other Side of Town

Talking faces
Wear the blues
Of singing faces

Thoughtful faces
Wear the blues
Of vocal faces

Laughing faces
Wear the blues
Of sad faces

Hopeful faces
Wear the blues
Of hopeless faces

Dressed up faces
Wear the blues
Of poverty faces

Sober faces
Wear the blues
Of drunken faces

Praying faces
Wear the blues
Of swearing faces

Love swoon faces
Wear the blues
Of hatestruck faces

Clean free faces
Wear the blues
Of jail faces

O side of town
Your sad faces
Are blues faces

Distance of a City

The travel was homespun.
By night
a lantern carrier beckoned us.

We covered the sea
like scattered wreckage. Villagers,
hidden workmen, came upon
stores, new tools, books.

This now is my town of regal presence.
Old stones hold up the flag, along
the river. Sounds here came

and modelled my echoes.
Steadily, wheels vibrate
dream-wheels. And a lot
of weariness faded here with sleep.

Well placed women saunter
in tiaras, in pearls, in finest fur.
Cleopatra's Needle stands.

Ethnic spirits in relics,
from South, from East,
from North and West, settle shapes
in the grandeur of rooms.

My belly is full of holes
like a tattered and flat hunting bag.
Bells ring: my belly is bottomless.
An alien sweat settles connections.

Ammunition I didn't make
explodes in my face
in a deathless no-return,
in a strange introduction bequeathed.

On an Afternoon Train from Purley to Victoria, 1955

Hello, she said and startled me.
Nice day. Nice day I agreed.
I am a Quaker she said and Sunday
I was moved in silence
to speak a poem loudly
for racial brotherhood.

I was thoughtful, then said
what poem came on like that?
One the moment inspired she said.
I was again thoughtful.

Inexplicably I saw
empty city streets lit dimly
in a day's first hours.
Alongside in darkness
was my father's big banana field.

Where are you from? she said.
Jamaica I said.
What part of Africa is Jamaica? she said.
Where Ireland is near Lapland I said.
Hard to see why you leave
such sunny country she said.
Snow falls elsewhere I said.
So sincere she was beautiful
as people sat down around us.

An Arrested Young Man's Reply

In shadows of backs turned
my muscles grew

In manhood unreachable
I absorbed

In noises of attack
my manhood overtook me

Captivated by what I have
I strolled the centre of town

In-a Brixtan Markit

I walk in-a Brixtan markit,
believin I a respectable man,
you know. An wha happn?

Policeman come straight up
an search mi bag!
Man — straight to mi.
Like them did a-wait fi mi.
Come search mi bag, man.

Fi mi bag!
An wha them si in deh?
Two piece a yam, a dasheen,
a han a banana, a piece a pork
an mi lates Bob Marley.

Man all a suddn I feel
mi head nah fi mi. This yah now
is when man kill somody, nah!

'Tony', I sey, 'hol on. Hol on,
Tony. Dohn shove. Dohn shove.
Dohn move neidda fis, tongue
nor emotion. Battn down, Tony.
Battn down.' An, man, Tony win.

Two Black Labourers on a London Building Site

Been a train crash.
   Wha'?
Yeh — tube crash.
   Who the driver?
Not a black man.
   Not a black man?
I check that firs'.
   Thank Almighty God.
'Bout thirty people dead.
   Thirty people dead?
Looks maybe more.
   Maybe more?
Maybe more.
   An' black man didn' drive?
No. Black man didn' drive.

From Lucy: Englan' a University

Darlin', you did ask a good question.
You know, an' I know, how white folks
can go on, 'bout they need a holiday,
an' manage get it 'cross
they 'adventurous', they 'curious',
an' land up in Africa
an' India an' China, etcetera, etcetera.

Well, darlin', Westindians come
get known as travellers for food
an' clothes: not folks who did long
to feel somewhere else,
long to touch snow with eyeball,
an' all curious too to see
horseback lords at home, at fireside,
maybe thinkin' how they did long
for company of bushman, an' did long
to take in sounds
an' feelin's of jungle places.

An' all in all, sweetheart, though
our move was blind date to Mother
Country, it look now it did carry
far-sightedness. An' I swear
heav'n did know something 'bout it.

Darlin', when real person get seen
in black figure, like when real
human get known in white figure,
is celebration, is teatime all roun',
or rum an' blackcurrant. An' still
I get surprise, when I see
somebody posh behavin' like
a hungrybelly in we distric' there.

An' real news this, me dear!
From distric' there, another
barefoot boy is makin' mark.
Fool-Fool Boy-Joe son come get degree,
education degree here in London:
an' Bareback-Buddy is teacher now.
Bareback teachin'. He teachin'
black an' white children mix together.
Leela, you see how what
you don' know is older than you?

Sweetheart, we mus' remember you know
how ol' people did say, 'Man is more
than a flock of birds.'

Parts and Wholeness

Ruminations

1
Warm hearts leap cold into fear.
Natives are compelling.
A face is like hillside to water.

Pool holds the scattered stars easy
as our heads hold magnificence loftily
as escapees in smiles of greeting.

Departure is footsteps unheard
in same swinging rounds
of palm trees and oaks.

2
Did skilled minutes work up
coconuts and strawberries?

How are leaves hanging feathers,
and scales and hands and hair?

I want the worksyard
with workshops

that coaxed up moonshape,
stone curves and wood and apple shine.

3
She stands with her summer spell.
Coal black she breaks into flame.

Transparent parading nothing
an abundance reveals and awakens.

Frequently exhausting,
she is inexhaustible.

Already the image: only
the acts of celebration we need.

A Companionship

I saw we moved together.
Often we were a bike, ridden
on two flat tyres.

Often we were a pair of fuses
well lit, hissing
but failing to feed the explosion.

Often we were stags
full of shots
bellowing to each other.

Often we were twin fields of words
like fallen apples
unharvested.

Often we were handshakes
that shook
damaged heads.

Often we were hidden stars
that never appeared
though sometimes nearly.

Ongoing Encounter

She said she called
in anguish, a wounder came
in a musical voice.

The blessing became a pain
manufacturer, the dazzling
face a death mask.

A rescuer came to her
loneliness, in a mood
lonelier than madness.

And desperate to forget,
she has a tormenting voice
that walks with her

unsettling her
in sullen helplessness,
preserving her like doom.

Parts and Wholeness 1

Not the same as left unpicked,
an absence which obliterates
tells the living did not occur:
there is no need of just rights or place,
there is no style that needs attention,
there is no depth to address,
there is no flowering to persuade.

Parts and Wholeness 2

This is a time I will not dread
the cunning of civilized faces,
the clean voices like machines,
the clear gestures
like condemning bugles.
We who look for inheritance lost
must search where the hope is wholeness.

This is a time I arouse
animation of shadows,
release uncertainty
in transformed openness.

This is a time I face a difference
for its substance of shades.
I absorb and not deface
with edges of schooling,
not allow good offices to spread killing
news, like golden girls out
showering couples with confetti.

This is a time I face up
to a light and a shade
for its glow or deeper reticence,
a taste for its bitter body,
for its salt or its sweetness
and a segment for the stone or flesh it is.

This is when I approach craggy face
of time and say, before I can settle
connections for synthesis, let us
let the waiting skeletons
restore their unique styles.

Stories by Bodyparts

1
I'm your breath-smell at waking.
I report your return from silence.
Use me in first kisses on children.
Use me in smiles on good-morning.

2
We are your eyes, your windows
on altitude and depth,
on faces hiding messages.
Sometimes we see
elevations unite levels.
Usually, horizons are fixed.
Blood boils us often
more than curved hips.
Bareness calls down hoods
yet you want to see the loving
next door: we wonder
what bird could be singing.
We wonder, to whom
could we expose we care.

3

We are your feet. We rubbed
away your baby tantrums
into leg-shapes of clockhands.
We stand at doorways:
no last abode anywhere.
Obsessed with wings, you make us
go about in traps.
We climb we fall.
We linger it hurts.
If you laugh or kiss, always
we are way at the end.
We dash for the bus, your
destination's eyes, high up
at its rear, laugh at us
and disappear. Then at last
we face outwards, like
clothes-irons rested
going colder and colder.
We see: no bath was ever able
to soak away our travels.

4

I am your hand. I hang
lightly or become a stone.
Agent of eyes and heart
I pull triggers.
I stroke up love.
I wrap up weapon.
Good cash or bad or its absence,
all arrive for pocketing.
And as this ladling palm, I take
a nail driven through me.
But like a leaf, I wave,
I shiver,
I shrivel,
I slip away.

I am Racism

I am the progress of tribalism.
I am civilized. I well know
the fixations of nonspecials.
I must protect pure and sane futures
and keep myself well walled.
You see I do deserve
the switch I press in others.
After all I am Racism.

I am the magic embodied before all eyes.
I am the most beautiful. I am
offended when nonspecials get seen
as my equal, since I already know
an ethnic difference is the sign
of nonspecial people. After all
I carry the supreme essence.
I have a position to uphold.
And I well know I am extolled
in secret by nonspecials.
You see I do deserve special rights.
After all I am Racism.

I am a royal household.
I live in an exclusive range.
Success of my special difference
displays me worldwide. And since
I have the ways to win, it helps
to underline the obvious.
You see I do deserve
the core of any country.
After all I am Racism.

I am the marbled column holding up
the sky. I must control to keep
entrusted gifts. I must wipe out
any sign to remove me, particularly
a sign from any nonspecial.
You see I do deserve the strength
I get reinforcements from.
After all I am Racism.

I am the church.
I must show how God is one-faced.
I am university.
I must show how to disguise
one dominance is absolute.
I am prime minister.
I must show how I keep my word
and defend goodness of my supporters.
I am the press.
I must recycle the people's confectionary.
I am the police.
I must bend laws in my defence.
You see I do deserve my slow
as well as my quick wits.
After all I am Racism.

I am the noontime in night-time,
the unique self, operated
in truth and trust. I'll be spoilt
if I change. I must defend to keep
the special success I am.
You see I do deserve
most wealth and ways and means.
After all I am Racism.

Looking for Houses

A desert captivates:
drawn to this place,
I look for houses of laughter
and schools of workmen.

I look for signposts of pain
weaved. I look for
a concordance of excursions.

A desert pulls quietly
into its stark destitution.
All echoes are my calls.

Elemental void fills
my head with dazzle,
with an absence of faces
with feeling, an absence of hands
to make acclamation.

Words fall from me like sweat,
other times like leaves.
Grapes grew here moments ago.
Palm trees were there.

Smells of food hung here in heat.
A stream was brilliant in the sand.
Giving coolness of a shadow stood.
Surely, here, a glory shone in everything!

So mazed by one path,
calling up another noon, night, dawn,
how was I drawn into this
hollow thought disembodied?

Inward Travel

Weighted by exhaustion
today, I have no eyes outward.

I have become a log
reverting through green.

I asend with rivers' faces
till again together I am

a drifter stilled, a raft
losing weight steadily

absorbing light's clarity
in expansive quiet.

Silence is hynoptist.
Magnitude washes

me in nothing. Then,
I see I have returned

in slow turn of a wheel
quickening.

Reconsidering

Reconsidering

Afterwards
the argument impressed me much.

I saw we suspect all life is good
and it is response
that makes it otherwise.
So to dispel any truth in that
we cajole each other to be hypocrites.

Hardest of all is how
not to be a hypocrite.
Everybody cajoles everybody
to be a hypocrite.

And especially
a chosen leader must first show
being the most adept at it.

New World Colonial Child

I arrive to doubtful connections,
to questionable paths,
to faces with obscure
disclosures, with reticent
voices, not clear why
my area is inaccessible,
my officials are not promoters.

Odd farthings drive the circles
bare, around the houses, like
goats tethered and forgotten.

I can't endure like my father.
I wait bowed. I wait
in rain-saturation,
in sunlight-dazzle.

Dark valleys and snow domes
are elsewhere. I am a piece
of disused gold mine, sometimes
a feather of shot game, other
times a seedbed of obsessions.

And making it is making.
Who isn't faced
to surrender money or knowhow,
surrender strength or bold,
reluctant or benevolent blood?

In the haughty handling
winners keep a military stance.
Avenues were never landscaped
for people blueprinted for rags.

A meeting has stayed
on a footing of war,
levels of weaponry unknown:
words misstate manoeuvres.
I hide. I admit
at best I am stubborn
like weeds on a path.

Absence of a choice has
a grasp of a slow death.
Absence of a hero makes men
headless, makes world-successes
work from failed retaliation.
And who shall expose
the virtues of difference?

Father's learning long taught
him, he's too lazy
to be man, too worthless
to be paid for work. He walks
like a loaded donkey,
unqualified to engage
essential listener. He knows
knowledge inflates a person
beyond little speeches in fragments.

I cannot assess my father.
I do not know what makes him
history. He's merely our
mystery of helplessness,
our languagemaster dumb
with forgetfulness, our
captain without compass.

And I cannot fathom
the people he's given me.
I still have to see if
our failures opened
inner doors of a meeting,
behind netting jaws, more firmly
than pain or profit.

I do not know my kinspeople
to be less than I know them,
yet judgements make me
feel they are less.

I do not know my losses,
I only sense them. I
do not know any licence
against me, it only seems so.
But a rope at my neck is
a shame I am born in
that I can't understand.

I hold on to a pride:
I own a map
underfoot. I own a king
and kingdom and robes
and rites I use.

On May 24 every year
I march, in fresh khaki shorts.
We wave the flag. I
with my slave scars march
and sing: Britons never,
never, shall be slaves.

In my dumbness I know
in our sky-wide gestures
gentle strengths arouse
a light in everyone.
How can I say what best
my heritage surrendered?

How can I know my voice
isn't the grunt of a pig,
isn't the squawking of a goose,
or the howling of wind?

A colony is a lair
of a country found. New lords
give names to people and places
and things and stamp them.
Equal terms would hold
a more-than-equal people
to a gunless robbery.

A colony has no resource
value for itself. A colony
never redeems itself with payment,
it merely receives.

A colony is given freedom,
when freedom has always been there.
A colony is given
Independence, when independence
can only be arrested.

Without plan or invitation,
like a season impels
I am charged to move.

I leave the encampment.
Like fresh awakening
I emerge round corners.
And again a different
weather is fierce
and I freeze-burn.

How shall anyone agree
a colony-native isn't
a colonised ghetto captive?
How shall we
clear away old orders?

The Presence

His appointments come on past dates
his arrivals are led to an interception
his handshakes greet fake supermen

His accent is the echo of loss
his steps make syllables missed
his movements are lacks choosing lacks

His money is trapped in places too locked
his answers are scattered in stones
his hauntings overload him with lives

His time is a house of questionmarks
his time accommodates void voices
his time makes mirages

His one love strips off masks and kills him
his dead loves make sure he can't dismiss them
his dead loves approach him day and night

His emptiness is all a wishful voice
his yearnings are all one cave of crooks
his busyness is all his invalidation

Letter to my Father from London

Over the horizon here
you say I told you
animals are groomed like babies
and shops hang wares
like a world of flame trees in bloom

Lambs and calves and pigs hang empty
and ships crowd the port

You say no one arrives back
for the breath once mixed becomes
an eternal entanglement

You say unreason eats up the youth
and rage defeats him

Elders cannot be heroes
when the young wakes up centrally
ragged or inflated on the world
and the ideal of leisure does
not mean a bushman's pocketless time

An enchanter has the face of cash
without sweat
and does not appear barefooted
bursting at elbows and bottom

He has the connections and craft
to claim the sun in gold
and the moon in diamond

You cannot measure the twig-man
image you launched before me
with bloated belly
with bulged eyes of famine
insistent from hoardings and walls
here on world highstreets
holding a bowl to every passerby

You still don't understand
how a victim is guilty as accomplice

Progress

1
Like dogs fed to kill, the 'best
educated' organized my skills.

I gave my ears my eyes
my hands my feet and feeling.

The 'best educated' drilled me
to give my strength to barriers
of old circles.

No need to touch other ways,
no new level to approach, our
developing turned back year after year.

Our journey a living place haunted,
famine keeping the well-fed distant,
war makes humans 'great'.

Our going repeating,
weapons being fangs and claws mutated,
our blank space in the head
accommodating bow and arrows
then guns then bombs,
the nuclear glow moves in and waits.

2
Early this morning I heard:
'In a way of touching other ways,
first, there is desire.
Even when other food is found,
it is hard to break up
the staging of the eaten lamb'.

Island Man

How I have watched you chop and ruffle
a tired face of land with hands
like dead roots, at full life
keeping back weeds
from a hillside patch.

How I see winds molest
your rags rotten with sweat,
see the sun paint you deeper and deeper
and suck your bosom dry.

Man grilled in sunken eyes,
minder of bush knowledge
the wind laughs at repeatedly,
knocking down your sticks,
beating up your few hills of yam,
I have watched you, seen you
stumble to shadows
reckoning insensibly
the season's and land's potential.

A new and clean voice burdens you:
your eyes hide from the meeting,
your humbled smile trembles with fear,
your arms are restless
like wild wings of a seized bird.

Unable to go any other way,
object of a landscape,
you lift weighted feet
with the memory of chains
in floods and arid ground:
your shoulder rocks a dirty bag,
your stumpy machete in your grasp.

Between bamboos and old palmtrees
you are in pursuit,
making me know I see a man
loaded in mind, but only
with the ways of his woods
that exclude him
and control him by compulsive tracks,
by strong sunrise
and its last rage of departure.

Luckless man of ceaseless attempts,
I have known you and watched you.
I watch you closer
now, watching myself.

New World Black Woman

Knowing its contempts and its loves
the world could not know you,
but you seemed ownerless
and your being enhanced fancies
and clamours emptied you
like inexhaustible land.
Often you held me in grief.

A rich valley found and ravaged
into open thoroughfare,
daily you were trampled
but never possessed: the art
we saw was violence in achievement.
Deeply you endured. It is good
sometimes you undressed beside me
saying an explicit 'yes'.

From your frugal presence,
greed would be reflected where
excess is an embarrassment.
Your space would emphasise
what is real and what is fantasy,
what is past and what ahead.

Low moving though not lowly
you turned your going into
a river's mysterious way:
desecration you suffer cleanses
desecrators, to see inclusive
creation. And slowly it emerges
earth stain of body is
a holy sign that is not land,
not mule, not any prey.

Your earth-and-sun thighs release
new movers globe-wide
as fresh landscapes of black.
Your freedom puts Europe in the red.
Old angels grow grubby wings.
It is good your giving is
essence your struggle distilled.

It's Me Man

I wouldn't be raven
  though dressed so
I wouldn't bleed my last
  though crushed
I wouldn't stay down
  though battered
I wouldn't be convinced
  though worst man
I wouldn't stay pieces
  though dissected
I wouldn't wear the crown
  though king of rubbish
I wouldn't stay dead
  though killed
I wouldn't stay dead
  though killed

Great Story

He drops out of sunlight,
voice like velvet, clean, unhurried,
says he's GREAT.

What makes GREAT, John asks him.
Knowhow, GREAT says.
Proof is the guts of GREAT.

His friends fall out of leaves.
More friends leap out of grass.
They burn John's house in a flaring fire.

John stands, unbelieving,
See? GREAT says. See change
in bricks to ashes.

See clean-up of fusion,
combustive excitement,
so well futuristic?

And there's more, GREAT says.
GREAT'S friends draw close to John.
GREAT'S friends hold John.

They break John's spine.
They crack John's skull.
They break John's legs and arms.

GREAT works on John.
GREAT works on a reputation.
Skills of GREAT rebuild John.

Reporters flock in with cameramen.
JOHN WALKS. JOHN IS REPAIRED.
O the extra excellence of GREAT.

The world is elated.
Relief absorbs the world.
The world enjoys relief.

I Spoke Severely

I spoke severely to my hands
for the chains
that contained my steps
    I go vigilant for the hold
    of invisible chains

I reproached my voice
for the silence
that silenced me
    I go vigilant for a loaded voice
    laden with silence

I scolded my being
for movements in the time
that arrested me
    I go vigilant for a full time
    that is not free time

I centred my mind
to dispel the surrender
that dominated me
    I go vigilant
    renewing my mind
    with constant eyes
    with constant eyes

Confession

I had a condition, she said.
I was born in England, you see.
Till last week, I was seventeen
years old. I've never seen
a Caribbean island, where my parents
came from. But I was born to know
black people had nothing. Black people
couldn't run their own countries,
couldn't take part in running the world.
Black people couldn't even run
a good two-people relationship.
They couldn't feed themselves, couldn't
make money, couldn't pass exams
and couldn't keep the law. And
black people couldn't get awards
on television. I asked my mother
why black people never achieved,
never explored, always got charity.
My mother said black people were cursed.
I knew.
I knew that.
I knew black people were cursed.
And I was one.
All the time I knew I was cursed.
Then going through a book on art
one night, a painting showed me
other people in struggle.
It showed me a different people like that.
Ragged, barefoot, hungry looking
they were in struggle.
I looked up.
The people needed: other people needed.
Or needed to remember their struggle.
Or even just to know
their need of struggle.
No. Not cursed.
Black people were not cursed.

New Reading Like Rebellion

We the new poems, we carry no roses
no snow or rhymes of rhetoric play.
A desperate breath, each
is a time stillborn, echoing.

We say, that bright new barriers came
drifting over seas into dreamtime
and settled over boundaries
where roots repeated, unsown,
and branches have their brain.

Struggle is what we carry:
deliveries of hot voices, and arousal
of own echoes' meanings, like
let my people go, let the offsprings'
happened innocence be right.

Mandingo faces, Congo faces — Yoruba,
Ashanti, Dahomey kingdoms — we show:
how they walked with the new names
of McKay, Reid, Dubois,
Mittelholzer, Bennett, De Lisser,
above memories of the Beaded
Crown and the Golden Stool.

We say, how these days some of the continentals
move bulldozed, and sunlight shows
the lovers fettered as outsiders,
on their own land's currency. We say,
in making disorder of a people
a State makes other ordered areas.
In having harvests of happiness out of them
the State empties a people of their happiness.

Archives papers, too, we are, you see,
inside stories that outside were demons:
all colours of different banners now.

We carry half the universe
not advertised as wholesome. We show
commercial skins not made extinct.
We outline debts too big to be acknowledged.
We say how everywhere a multitude,
or the global specks, walk about
obsessed, chanting no! no! no!

We say how they say they have saved
themselves, and how they meet any time
any place and break out in dance and hurrahs
with drinks in hand, over not all
being bones underwater and underground.

We say firstclass brains swept
the libraries, dusted seats
of the legislature
and polished dignitaries like gold gods.

And becoming both the tract and the testimony,
from prison to the PM's palace,
they were central in the new messiah's
motorcade, in a nation's jubilation
through shanty towns
and through the immaculate avenues.

We show the ongoing aftermaths:
the ancestors against missionaries,
the country boys coming to town,
seeing, smiling, and believing in
the cessation of death like fledglings.

And in the renunciation with open arms,
a road goes through The Palmwine Drinkard,
The River Between is crossed,
Things Fall Apart:
Arrow of God is Houseboy,
is The Spook Who Sat At The Door,
is The Invisible Man,
is The Fire Next Time:
In The Castle Of My Skin
Black Boy
Song Of Solomon
The Arrivants
Where The Hills Were Joyful Together
In A Green Night
In Wings Of The Morning.

Then comes this news,
this supreme soldiers' time,
unseating leader after leader,
and the flights from The Islands
to have the money away from attempts
at abundance in the villages.

How then shall the fractures be the aesthete?
How shall there be no prop-up by gun
or 'weed', or a disguised waiting
on words or moves from the boss?
How shall the colourful abundance assemble?

Benediction

Thanks to the ear
that someone may hear

Thanks to seeing
that someone may see

Thanks to feeling
that someone may feel

Thanks to touch
that one may be touched

Thanks to flowering of white moon
and spreading shawl of black night
holding villages and cities together

Insatiable Mover

Echoes pull echoes drive him:
  he'll have open smiles of green walks.

He insists from a fleshy womb and stands:
  he'll outdo the height of trees.

He wants a body agile and ageless:
  he feeds on flesh.

He wants the trees' deep singing:
  he settles himself within stripped wood.

He wants the splendour of petals:
  he wraps himself in colourful cloths.

He'll have the sun's dazzling scope:
  he hunts the ground and blood for gold.

He dreams of a devouring love:
  junction of a woman's thighs arrest him.

He'll have disembodied thoughts:
  words in print feed his insatiable eye.

He'll have the insides of mysteries:
  he breaks up air he exposes embryos.

He longs to manage an earthquake's might:
  he makes bombs make cities collapse.

He longs to escape from moods of melancholy:
  he lets music entrance him.

He longs to absorb length and breadth of the globe:
  he builds crafts and travels.

He longs to top up his height with stars:
  he reaches for the sky in a rocket.

He longs to draw on full depth:
  he goes down he sits in a pothole.

He longs for time gone and time to come:
    he sings he dances he laments.

He longs to confirm he grasps and wins:
    he flashes light he outpours darkness.

He'll have that answer that is absolute:
    he finds new ways to make disorder.

He'll have his voice collect all feelings:
    echoes pull echoes drive him.

He longs to beat all haunting:
    cessation takes over his limbs his lips
    his eyes his ears and stops his being.

Thinking Back on Yard Time

Thinking Back on Yard Time

We swim in the mooneye.
The girls brown breasts float.
Sea sways against sandbanks.

We all frogkick water.
Palm trees stand there watching
with limbs dark like our crowd.

We porpoise-dive, we rise,
we dog-shake water from our heads.
Somebody swims on somebody.

We laugh, we dry ourselves.
Sea-rolling makes thunder
around coast walls of cliffs.

Noise at Square is rum-talk
from the sweaty rum bar
without one woman's word.

Skylarking, in our seizure,
in youthful bantering,
we are lost in togetherness.

Our road isn't dark tonight.
Trees — mango, breadfruit — all,
only make own shapely shadow.

Moon lights up pastureland.
Cows, jackass, all, graze quietly.
We are the cackling party.

Memory

Dawn hangs a crimson dusk
of all the Flame Trees in bloom
with all the hibiscus. Bare feet go
collecting animals in heavy dew.

Morning delivers new sun.
Smoke rises from little kitchens.
Every passing person calls out
'Good morning!'

Eyes go drowsy in blazing afternoon.
Golden wasps shimmer.
Clothes collect summer smells.

Swift brown bats litter dusk.
Anancy weaves in log cottages.
Grandma's knowing ripens.

Night ferments mango-walk.
The caring night pouches
busy sea and static hills
together with flatland palm trees.

Anancy: spider hero of Caribbean folk tales

Nana Krishie the Midwife

So keen on me those old eyes
the tracked black face
flowed with light

The tongue and gum ladled
stubborn words remembering
how I the boy child had knocked
thirty years before and hustled her
to come to the little cottage

Come with owl's wisdom and red
calico bag of tricks
to end labour: snap
and smack a newborn to cry

And now she looked at me surprised
and not at all surprised I had
come back from abroad
looking in a widened range
out of miracles she used and knew
time had discredited

For her ancestor's knack
her tabooed secrets now worked
in books of others
as ancient practices

Dreaming in her illiterate life
I felt the faltering tones
her startling shivered voice
thanking God
for showing me ever so well

Black Statue

(Paul Bogle, Jamaica)

New form on my eyes now:
the embalmed is unclothed
a figure of flame.
Cut-down man is mounted.

Refaced man reveals
I echoed heroes
I never felt a friend.
Go easy on me.

We were similar men
of the same allegiance,
the same yearnings
to unfetter a finite face.

In days and times
chained together
we interchanged longing.

Torchlights have marked
your angles among
chains and wrecks.

Impulses you store consume
too much: let me
measure my pace.
But hold me a while.

Hold me in
failed sinews and blood
and burning thoughts.

You moved
for my release,
you died.

Erected from oblivion
you are awesome here
how the sun reveals you
signing me up a dazzler

with trees and flowers
for sores and curses,
in a place
starting up a world.

Reclamation

Suspended and choiceless
I grew in abeyance.
Memories mocked me.

Old sun-scented beds were haunted
realms, of gestures unfinished.
My path was a tunnel
ending into nighttime.

Nature obeyed no black man:
all the lands, oceans and space.

Black men swopped gold for trinkets.
They played hide-and-seek with ghosts
and raised up dust for rain.

Yet there was a knowing I was
marooned from. And I didn't
know what or how or why.
My sanctuary held no truth
for I could not enter it.

Voices in me grew beyond me.
Voices in me grew to pentecostal bedlam.
I knew I must retrace a travel.

I knew I must go back
through groans and griefs
through putrefaction
through shit of ships' bowels
through staring eyes and tears unwept.

I must go back through a march from home
through graves at sea without goodbye.
I must go back
through friendless arrivals
through beautiful bewildering faces
through change of name
through loss of tongue
through loss of face
through exclusion and sweating
for mountains of money
to the grave of a slave
through people fixed in a ship like wood
in long listless days and days
over the swells of indifferent seas.

I must absorb you
Middle Passage,
must refine you, must
distil a journey, like any,
into an innocent voice.

Here I am.
Here I am, where
at beginning one sun
daubed and brushed me
in silence, and I became
obsessed lover of the dance.

And here are these eyes.
All unfathomable eyes surround me
in motherland.

Unreachable time has turned
familiar voices strange,
but kept every face my own.
And none is my grandmother.
None is my grandfather.
None is my known cousin.

But called or uncalled, these are names
that are musical instruments
that announce death and life,
announce trespasses and hauntings,
announce a welcome and redemption.

Reinitiate,
rededicate,
o all of me,
all of our dead
dumped at sea unloved,
and those given the destination
to a life they never owned.

Approach me.
Approach me
drums, whistles, chants.
Approach me.
Reconsecrate the days that carry me.

I hear riddles.
I hear proverbs
mixed in drumbeat.
I hear the time of day-one.

Exorcise me.
Exorcise a castaway
in hypnotic rites.

I am a sacred place.
I am the beginning of my ancestors.
I open the way as first and close it as last.
I am a flaming stone.
I am a tree never to be cut
and a word reserved for ceremony.

My tongue chisels and
my tongue churns old words. I'm
rolled back from sand like ocean waves
that rise to begin again.
I sing an old song like a first song.

Release is an unworded pledge.
Impulses poise my body.

We dance.
We dance in dust.
We dance. We dance. We dance.

With found faces and drumming,
with found faces and drumming
I'm new spirit out of skin.

I'm new spirit out of skin,
with found faces and drumming . . .

A Show of Hidden Territory

I was no more afraid
of the people's beauty.
Their angelic masks began to fall
away, like loosened bark.

I was amazed.
The music of voices joined masses.
Even in deluding,
some sounds could not be static.

Rastaman

See him there,
understand
he is maker of religion.

He is repudiator.
He emphasizes
image of God
in an indelible presence.

Understand,
he restages
a stubborn black man's count.

A bush voice booms
there's no model people,
there's only another face,
another variation.

Understand,
because it is good and pleasant
for brethren to dwell together

his dreadlocks assemble
a garb and warrior gear.
And in his wine of weed
in a drumming in Psalm 133

he stops
Babylon
trampling him.

Understand,
it's no rebel head he rests with:
it's an open treetop
for revelation beyond Reggae.

He'll reclaim
creation
in man walking Westindian.

Goodmornin Brodda Rasta

Good-days wash yu mi brodda
a-mek peace possess yu
an love enlightn yu
a-mek yu givin be good
an yu evermore be everybody
a-mek Allness affec yu always
an yu meetn of eye to eye be vision
an all yu word dem be word of wondament

Approach and Response

Approach and Response

Now you have excited her.
You have started the sound
of the abrasive wheel:
that heart, so tightly tied
with pain and tucked away.

You are still far far
from a welcome touch in this house
and the one chair that holds
a wispy woman through
most nights. But you are
guilty of arousal.

You have knocked a door.
You have shaken up death again.
You have rattled account demands
impossible to settle.

Our Love Challenge

(for Mary)

Maybe designs for conflict worked me
when I exploded on the stage
with slave anger
thoughtless you'd walk out wounded

So full in time I let out
my raging hungry dogs
who threatened me

You take it I accuse you
of seizing children's keep
with whip with cruel lordship
you say I invest in vengeance
I storm you with ancestors' sins

You cut all lines to me
you go to old fires behind barriers
I like an outlaw withdraw

But alone again leaving
tiger and lion interlocked
we go slowly

We exhaust suppositions
we become this new area
between killing sun and killing ice

A Woman Reflects

A stepping back not offered,
time folds up its tracks
with actions of light and darkness
into layers of memory.

And with the travel not noticed,
with its drama scarcely appraised,
at this place today,
a beginning is staged.

A changed past becomes
new brute life
in a formed bridge to food
and the outlook.

Behind are the burnt out
and closed pathways,
the mummified things
with some not quite lifeless.

Then towards one's restful dawn,
a stranger like gnarled sticks comes
and shouts up the household,
announces Good Morning and sits.

And behind a hassle today
the heirloom ring is sold
towards leper-care. Old pictures
and a tapestry go for skeletons.

Other women sit in doorways.
Some encircle offices. All say
they'll release armies
from hypnotized killings.

Yet, old flavour of teatime
is sadly mixed with new
prayer mutterings on the air
and government cannot act.

What conviction could have been seen
as pain circulated? Could
my coat have stayed mink's,
and not a lady's best garment?

Commune Bookseller

At the Festival of Health
at her literature stall,
in sandals, T-shirt and jeans
with braces and cap, she faces you
with concentrating eyes.

She receives your interest
solemnly like a doctor,
till steadily her persuasion
is evangelistic, and you are
a fixed listener.

Are you a ready vehicle
that could take on
your regular woman's goodnight kiss
with her footsteps away
to a group brother's bed?

Have you a development
that would allow you to tell
your lady, another group sister
wanted you all night, because
both needed the loving?

Could you give your weekly cash
to everybody's budget?
Could you pool your personal rights,
be a communal dad, be mum
by group consent, to grow with giving

and be non-possessor, yet
having everyone and everything?
    And, you have to go: her words
are blank seeds in you,
awaiting their programme.

Young Woman Reassesses

I am told I enclose
the drives of a landscape
like a bunch of black grapes.

Just to stand at Oval Station
I get eyed like roast meat or get touched.
Any wonder I'm now a mask of stone?

I am told I conceal
warm rain and earth
in a fleshly fruit.

Hitchhiking I find friendly enemies
and have to take on
unknown highways.

Safe at last I go flying
with my talented and beautiful black boy.
I crash. I limp to the world for sympathy.

I am told, my eyes
and lips and all
unlock moments.

I release my international streak.
I go friendly.
My Rastaman all but kills me.

I bead my hair.
I tint my fingernails.
I colour my face.

I put on my best discoverable underwear
only to see my body is not
best friend to advise me.

I give myself to beauty contests.
I shine. I glitter. I bubble.
Hot baths become best friend.

Any wonder I want
a new style
for country and for town?

Absorbed deeply in a landscape I learn
I have a mistaken voice
calling for company.

Thinkin Loud-Loud

Yu sen fo we to Englan, she sey.
Yu buy de house.
Yu buy de car.
We inside dohn roll fo food.
I expectin yu numba 6 child.
Why yu beatin out yu brain on books
wha tight like a rockstone?
Find teacher. Find teacher, she sey.

Gal, I sey, fifty year I walkin earth:
ow can I mek a teacher wise
abc still a-puzzle me?
Ow can I show we own boy an girl dem
words on me eye put up high wall?

She sey, yu sign yu name wid X.
Yu show no paperwork, but
yu av yu workins
wid pencil an paper in yu head.

Gal, I sey, dat worded page
is a spread of dead tings: insect dem
wha stares at me
doing notn sayin notn
but turn dark night
an bodda me an bodda me
fo dat time I hear print a-talk
like voices of we children.

Speech for Alternative Creation

Let us recreate all things in own image.
Let us make a new beginning.
Let us remove night, dawn, dusk,
remove black thunder, leave lightning,
dismiss Dark November, leave
all eyes on noon, that dazzle
of summer, our white heat of days.

Let us remove the Dark Continent,
and be humane to drifters.
Let us build up a government Blackmale list
and a Blacklist of Blacklegs
for a kind and natural exile use.
Let there be no tar to dip the brush in,
though twin adjectives Big and Black can stay,
to be own reliable terror-image.

Let us remember: Blackmarks have a way
of coming to walk like children.
Let us remember this:
each boy our own, each girl,
is our own support garrison.

We shall let it be law, that anything
except immaculate conception
is Miscegenation, that no
white lady shall ride a black stallion,
that all ravenhaired people be made
whitehaired, all black Bibles be cancelled
for white ones, and that pure milk
shall replace all coffee drinking.

We shall get every Blacksheep into
a straight scapegoat, and be clear
that no Blackguard can ever be fair.

We shall erase, correct out, every printed word,
have pure pages to show decent minds.
Even trains shall have no tunnels to go
through. We shall make all our roads into
journeys on silver. And our science
shall treat all brown earth till changed,
like beauty of clinical cottonwool.

After all, a non-white thing is a non-thing;
Might as well will it away.

Let us deport devildom. Let us have
that dreamed-of constant world, mirroring
the sun with white lands and houses and walls.
Let us make it all a marvel of a moonshine.

The Meeting of Potentials

The bellies lined with waiting
like tissues hardened, again
our desperate philosophers are here.

In this room they'll mesmerize
barriers, levels and distances
that eat up their time,
that provide their unreadiness
and static framework
their children go around.

They are carriers of waiting
and rage, over damages
unaccountable, over their gardens
rotten with junk, over
journeys they cannot make
and rooms they cannot furnish.

Every ticking head knows
a leader is hidden there.
Nobody admits the parrying
of words at first is in fact
first shots of war.

Then business being too defensive
and too attacking, is heightened
into a roomful of punchups,
into tables and chairs left
at knocked angles
with nothing more to say.

The cleaner comes.
The cleaner whistles jauntily.
The cleaner straightens the place.

Last Freedom of Martin Luther King

Who is here without grief?
Who cannot share
our fruit of death fallen before us?

In the secrets air and clay emphasize in blood,
winding rivers do not mourn
in processions of children, men, women.
We have roads endlessly.

Only last night I looked on your living face;
a man unafraid in deadly shadows.
Today your exit dims a world.

You who returned no blows,
you who would undamn abundance,
you who would see enslaver and slave released,
you who would let illusion fade,
you who would arouse new day from wreckage,
you who would cut roads of diamond in city jungles,
you who in all this make us mourn.

O enemies are defined,
madness is courage.
A dead past animates us,
our ghosts insist.

And dream of brotherhood fixed your eyes.
Agonies of silent people lit your thoughts.
Citizens' birthrights must not be bargained.
"We Shall Overcome" moved, like moving
hives of bees behind you,
along the longest highways.

You linked people whose visions
were intolerable. You engaged the trapped
and hungry and quickened them, releasing
a new selfknowing like fire.
Yet in the middle of a beginning
we mourn you here and our innocence.

African Holiday

(for Nelson Mandela)

A strange wait
how leaders rot
swishing flies

A strange decay
of presence
walled up with words

A strange cancellation
of arrival
and promises

A strange mission –
tomorrow's news gone
like rolling waves arrested

A strange departure
of voices
a place developed

A strange reversal –
offering death
life's prime giving

Nelson Mandela: South African leader of African National Congress. Sentenced to life imprisonment in 1964.

Calabash Tree

You drew me to come
to you, Calabash Tree,
short trunked with long and leafy whips
at foothills here in the bush.

Your shape is
a confusion of fountain jets:
your lines make
an extraordinary wild face.

Your loose open top receives draperies
of sunlight.
I move and rub
your thin flakes of bark.

I climb up easily and sit: I know
streaked with purple your blooms
of yellow surprise with stink.

Scattered on the network of branches
or stuck to your trunk
green fruits are solitary.

On the ground some are rotting skulls.
And tied young they would have flowed
into pearshape
or oblong or other forms
for maracas, water gourds, ornaments.

You make me wonder.
You let it known your parts are fibred
for ribs of boats, tool handles,
cattle yokes or saddle trees.

Did you start out to surrender
at arrival, to become scattered
pieces, on the sea, in men's hands
and backs of animals?

I wonder.
I wonder how you first
received your impetus
to show the sun
its transformation.

New Empire

A blending of the beasts
shortens their fangs. The sun
comes over a trek reversed

and drilled religiously to win,
like squirrels will have
an assembly of nuts, the sovereign states
from high seas annexe nations

and become captain's wheel ornament
on wall of the Free House
and pieces of garden fence.

A commitment a kind of completeness works
an empire into museums.
Mosques and temples come
around a cathedral.

Sounds of brass and oboes
and violins meet tablas and steel pans,
all to a new song and dance.

The Coming of Yams and Mangoes and Mountain Honey

Handfuls hold hidden sunset
stuffing up bags
and filling up the London baskets.
Caribbean hills have moved and come.

Sun's alphabet drops out of branches.
Coconuts are big brown Os,
pimentoberries little ones.
Open up papaw like pumpkin you get
the brightness of macaw.

Breadfruit a green football,
congo-peas like tawny pearls,
mango soaked in sunrise,
avocado is a fleshy green.

Colours of sun, stalled in groups,
make market a busy meeting.
The sweetnesses of summer settle smells.

Mints and onions quarrel.
Nutmeg and orange and cinnamon hug
themselves in sun-perfume.

Some of the round bodies shown off
have grown into long shapes.
Others grew fisty and knobbled.
Jars hold black molasses like honey.

And yams the loaves
of earth's big bellies and sun,
plantains too huge to be bananas,
melons too smooth to be pineapples —
chocho, okra, sweetsop, soursop, sorrel —
all are sun flavoured geniuses.

Nights once lit the growing lots
with fields of squinting kitibus.
Winds polished some of the skins cool but warm
when sun drew stripes on fish.

But, here, you won't have a topseat cooing
in peppers, won't hear the nightingale's
notes mixed with lime juice.

Red buses pass for donkeys now.
Posters of pop stars hang by.

Caribbean hills have moved
and come to London
with whole words of the elements.
Just take them and give them
to children, to parents and the old folks.

kitibu: the click-beetle, or firefly, with two luminous spots that squint light in the dark